Tatiane Rocha Carias
Tais Arthur Correa

Introducing the theme of medicines as a mediator in chemistry teaching

Tatiane Rocha Carias
Tais Arthur Correa

Introducing the theme of medicines as a mediator in chemistry teaching

Experiments with medicines to identify organic compounds

Imprint

Any brand names and product names mentioned in this book are subject to trademark, brand or patent protection and are trademarks or registered trademarks of their respective holders. The use of brand names, product names, common names, trade names, product descriptions etc. even without a particular marking in this work is in no way to be construed to mean that such names may be regarded as unrestricted in respect of trademark and brand protection legislation and could thus be used by anyone.

Cover image: www.ingimage.com

This book is a translation from the original published under ISBN 978-3-330-74510-0.

Publisher:
Sciencia Scripts
is a trademark of
Dodo Books Indian Ocean Ltd. and OmniScriptum S.R.L publishing group

120 High Road, East Finchley, London, N2 9ED, United Kingdom
Str. Armeneasca 28/1, office 1, Chisinau MD-2012, Republic of Moldova, Europe
Printed at: see last page
ISBN: 978-620-7-89380-5

Table of contents:

I dedicate this work to God, for

being essential in my life, the

author of my destiny!

ACKNOWLEDGMENTS

To God for giving me the health and strength to overcome difficulties.

To UEMG, its teaching staff, management and administration, who gave me the opportunity to glimpse a higher horizon.

To my advisor, Tais Arthur Corrêa, for all her support, corrections and encouragement.

To my parents and brother for all their unconditional support.

To FAPEMIG for the research project grant that made this work possible.

To the Senador Levindo Coelho State School and to teachers Fernanda and Cristina for making their classes available.

To the 3rd grade students for their special participation in making the results possible.

I would like to thank the examining board, Tatiane Teixeira Tavares and Aline Aparecida Angelo, for accepting the invitation and contributing to this work.

I would like to thank my friend Leynara for always being there to help me run my classes.

Finally, to everyone who made this work possible!

CARIAS, Tatiane da Rocha. **Contextualized application of the theme of medicines as a mediator in the teaching of organic chemistry.** 2016. Course Conclusion Work (Graduation). Chemistry Degree Course, Ubá, MG, 2016.

CONTEXTUALIZED TEACHING OF MEDICINES AS A MEDIATOR IN THE TEACHING OF ORGANIC CHEMISTRY

SUMMARY

The Medicines theme represents an interesting proposal for teaching chemistry in terms of incorporating contemporary and social themes, addressing CTSA (Science, Technology, Society and Environment) aspects into teachers' pedagogical practices. The work involves contextualizing the teaching of organic chemistry through the theme of Medicines, self-medication and the disposal of pharmaceutical products, using different teaching methodologies. It was developed in partnership with the E. E. Senador Levindo Coelho, located in the city of Ubá-MG, with approximately 200 students enrolled in the 3rd grade in the morning and afternoon shifts. The work involved a preliminary assessment of how the subject of organic fungi is dealt with in the school in question, together with a superficial analysis of the textbook. Expositive and experimental classes were applied, developing the proposed theme through problematization strategies with subjects related to the students' daily lives. The majority of organic chemistry classes are taught in the traditional way, with little relation to the students' daily lives and without the use of teaching resources such as more dynamic and experimental classes. Through the use of differentiated resources such as visual tools, the application of contextualized practical lessons and the analysis of the data obtained, there was greater assimilation of the organic chemistry content. In view of the results obtained before and after the methodologies were applied, it can be concluded that there was a greater understanding of the subject, making the lessons more attractive with the use of the Medicines theme, contributing to the students' critical and civic education.

Keywords: Context, organic fungi, medicines.

Chapter 1

1 INTRODUCTION

Chemistry teaching should help students understand the world around them. The school has the role of training and preparing citizens to act consciously in society, addressing issues that can be contextualized with reality (PAZINATO *et al.*, 2012).

Chemistry content becomes more interesting and enjoyable when students are motivated through differentiated strategies that promote interactions with their daily lives. The use of contextualized lectures and experiments helps with learning as they are methodologies that encourage the visualization of reaction phenomena. The search for understanding through the use of visual materials and practical lessons with substances that the students are familiar with can guarantee better assimilation of chemistry content when approached in a contextualized way (ALBA; SALGADO; PINO; 2013).

Contextualizing subjects related to society is a practice designed to provide critical and social training, using subjects that can revive the importance of education in society and the environment. Among the different themes, the subject of medicines can be an efficient study tool to promote a relationship between students' everyday lives and the subject of organic chemistry. This approach can help contextualize the content covered, make it easier to memorize functional groups, relate the possible structure/activity of medicines through the chemical structures of their molecules, their positive and negative effects on our bodies, or even the environmental impact caused by their incorrect disposal in the environment (ZUIN, FREITAS, 2008; SANTOS, 2007).

In view of the above, the approach to self-medication and the conscious disposal of medicines, from an educational perspective, becomes an interesting proposal for teaching chemistry in terms of incorporating contemporary themes, in a contextualized way, into the pedagogical practices of teachers, giving meaning to the contents of Organic Chemistry through themes related to Science, Technology, Society and the Environment (CTSA), directly influencing people's quality of life, as well as contributing directly to the development of the social protagonism of the student community.

The purpose of this work is to highlight the use of the theme Medicines as a mediator in the teaching of organic functions, using drugs that are present in the students' daily lives through the application of expository and experimental classes that can arouse interest in learning this content, as well as promoting a relationship with issues pertinent to society such as self-medication, excessive consumption and disposal of medicines.

Chapter 2

2 OBJECTIVES

2.1 General Objective

Using the theme of Medicines as a mediating theme for teaching Organic Chemistry, making a correlation between self-medication, excessive consumption, disposal, environmental awareness and the study of organic fungi, enabling greater learning of the subject, as well as arousing the interest of teachers, high school students and undergraduate chemistry students in scientific knowledge by solving problem situations related to their daily lives.

2.2 Specific objectives

- To find out how teachers develop the content of organic fungi through structured questionnaires;
- To understand the students' affinity with the subject of chemistry, their ability to relate everyday substances and their knowledge of the contents of organic chemistry through the application of a preliminary questionnaire;
- To analyze how the content of organic fungi is laid out in the chemistry textbook used at the school;
- Lectures and experiments relating medicines to their functional groups, then questionnaires to assess how the practice contributed to the students' learning and their representations of the appropriate use of medicines;
- Encouraging effective student participation, giving them opportunities to put the knowledge they have acquired into practice and stimulating enthusiasm for the subject.

Chapter 3

3 LITERATURE REVIEW

3.1 Teaching Chemistry

Chemistry plays an important role in technology and society by contributing to the development of all branches of science, as there is no material that is beyond its reach. The teaching of chemistry should stimulate and exercise attitudes that favor interpersonal relationships, relationships with society and relationships with the environment. An appropriate conceptual program of different approaches should be promoted, linking theoretical aspects with their phenomena and representations. One of the great difficulties in learning is often due to the traditional and mechanized way in which content is approached. They are often taught without establishing connections with reality (CARNIEL, 2013).

Students find it difficult to relate applied theory to the reality they experience. Media resources in lectures can be used to make this connection between applied content and important current issues. Visualization has become a facilitator and has been developed in the learning process for the development and application of science, which begins with the presentation of figures, pictures and graphics. Currently, equipment such as television and computers are used as a methodology to help explain and grasp content (RAUPP, SERRANO, MARTINS, 2008).

In addition to lectures, another methodology that involves visuals is experimental lessons. Observing phenomena promotes students' interest in the content and arouses their curiosity, encouraging scientific thinking through everyday situations. In order to achieve meaningful learning, it is necessary to link theory and practice, creating conditions for students to understand the subject through the application of facilitating methodologies that promote a relationship with issues experienced by the students (BUENO, 2007; REGINALDO, 2012).

According to Vygotsky (1989), practical lessons encourage curiosity, determination and decision-making, improving the development of linguistic and intellectual capacity. In addition, experimental lessons help with student interaction when carried out in groups, stimulating teamwork. The assimilation of content is based on social relationships mediated by symbolic systems. Practical lessons can be used as a means of mediation, establishing a link between the objective and subjective worlds, enabling the individual to know and learn.

In view of the advantages of practical lessons, attention must be paid to how they are prepared and planned in order to establish the best application strategies. The teacher must know the meaning of the content, enabling an approach that can contribute to scientific knowledge through relationships with the students' daily lives, developing strategies that can link theoretical phenomena to experiments, where through observation, the student will be able to elaborate explanations for scientific knowledge through an investigative methodology (MEDEIROS *et al.,* 2013).

Investigative experimentation establishes the construction of science, when students are faced with a problem situation that allows them to build their own knowledge. With their culture and perception of the world, students are in a position to be part of knowledge, they just need to create possibilities by inserting mediating instruments that can develop their skills, abilities and creativity (OLIVEIRA, 2012; FERREIRA *et al.,* 2010).

Understanding chemistry teaching using everyday resources makes it possible to develop a critical view of the world, enabling students to participate in situations that contribute to issues related to society. According to Paulo Freire (2014), education is an act of knowledge, not simply the transfer of knowledge, because the subject has historical conditions that favor learning.

Understanding the real situation of the world in which man is inserted must be introduced into the school environment, enabling a dialog between students and teachers that allows them to learn about new ways of understanding current problems by correlating reality with the disciplines.

In the classroom, knowledge should be actively grasped by the student, being an environment for socializing knowledge and acquiring new experiences, with activities that provide not only growth and learning in concepts and technical applications, but also in the application of these concepts in out-of-school situations. To this end, the teaching strategies should make the teaching-learning methodology more dynamic, in order to turn the learning environment into a place where experiences and knowledge are constantly exchanged and to enable a link to be made between the knowledge taught and the student's everyday life (CARNIEL, 2013).

In most Brazilian schools, teachers use very abstract and specific language. Chemistry content is taught in a way that is disconnected from the students' reality. This way of teaching and assessing requires students to memorize concepts. Of course, the consequences of this method are negative, because after a few days, students forget the concepts they have memorized, as they are not very meaningful for them to grasp. Perhaps the greatest negative consequence is resistance to wanting to learn, understand and live with chemistry (ALVES FILHO; RICHETTI, 2014).

For Lev Vygotsky (1987, 1988), cognitive development, i.e. the process of the emergence of the ability to think and understand, cannot be understood without reference to the social, historical and cultural context in which it occurs. There are two types of mediating elements: the instruments that create a relationship between the individual and the world, and the signs that correspond to psychological activity through cultural resources that make it possible to stimulate social relationships. This relationship can be used as a means of facilitating the development of content when helped to solve problems.

However, teachers still use a practice based on behaviorism, where teaching is transmitted through stimuli, responses and efforts, without much meaning. For Guimaraes (2009), meaningful learning is carried out by the subject themselves, where new information is related to relevant aspects and expressed in different ways to acquire and store the vast amount of ideas and information represented in any field of knowledge.

With this, the understanding of this knowledge should be broadened to include other technological, social and environmental issues from a contextualization perspective, enabling the development of scientific knowledge and students' civic education with approaches that make it possible to work on real problems that are part of students' daily lives (RIBEIRO, GENOVESE, COLHERINHAS, 2011; MARCONDES, 2009).

3.2 Contextualization in the CTSA approach

The approach to content from a Science, Technology, Society and Environment (CTSA) perspective is currently being proposed in the Curriculum Guidelines as a way of promoting student interest and enabling the development of knowledge of science in decision-making, improving students' critical thinking (FERNANDES, 2011).

The process of developing citizenship can be achieved by learning scientific, technological, social and environmental concepts. Whether they are natural or constructed, they can be used in the teaching process, contributing to the development of chemistry concepts (REBELLO *et al*, 2012).

The development of science can be attributed to teaching and learning situations based on solving problems present in our daily lives. Discussing issues related to CTSA provides an understanding of phenomena that can be debated in the current scenario (ZUIN; FREITAS, 2008).

Contextualization is used as a way of relating school content to issues relevant to society, taking into account not only the application of concepts, but also the development of strategies that raise concern about social impacts (MARCONDES, *et al* 2009).

The goal of teaching science through the inclusion of topics that are present in students' daily lives gains importance in investigative attitudes, because contextualizing the student's search for new knowledge leads them to reflect on their understanding of certain subjects, making it an important tool for constructing meaning (WARTHA, 2013).

Medicines are made up of various chemical substances that have numerous organic functions in their structure. Each organic fungus has an atom or group of atoms that characterizes the function to which the compound belongs. These atoms or groups of atoms are called functional groups (PAZINATO *et al.*, 2012).

Contextualization through the use of medicines to teach the content of organic fungi can help teachers correlate with the subject of organic chemistry. As well as being conceptually rich, because it allows teachers to work with molecules that have various functional groups in their structure, this theme of medicines also contributes to students' civic education (PAZINATO *et al.*, 2012).

In chemistry classes, it's not always easy to find a topic that establishes links between everyday life and the concepts being taught. The subject of medicines can be approached with the aim of contextualizing the study of organic chemistry, as well as promoting issues related to their correct use and disposal (FARY *et al.*, 2012).

3.3 Self-medication and drug disposal

Medicines are special products, made under strict technical control, to meet the specifications laid down by ANVISA. Their purpose is to diagnose, prevent, cure diseases or relieve their symptoms. This effect is related to their composition, which includes organic molecules with functions that have an active ingredient capable of reacting due to their specific properties (SILVA; FAGUNDES, 2010).

Today, medicines are synthesized in the pharmaceutical industry, which is one of the most important private economic activities in the globalized world. These industries are part of the so-called medical-industrial complex, because in addition to their important role in producing medicines, their main function is related to the marketing of this product (TRIBESS JUNIOR; ZANCANARO, 2013).

Encouragement from advertisements and the need for practicality have become increasingly present in people's lives, which ends up generating high consumption. There are several reasons why medicines are used incorrectly. Firstly, lack of time and easy access to pharmacies lead people to self-medicate.

For the lay consumer, medicine would be the magical possibility in the desire to acquire health through technology and science, made available and represented by small bottles and/or a few pills (RICHETTI, 2008; RICHETTI; ALVES FILHO, 2009).

In Brazil, self-medication is a practice that spans generations, from home use of medicinal plants to industrialized products. Its denomination includes the various ways in which a person decides which medicine, how and when to take it in order to alleviate symptoms or in search of a cure for their illness (ALVES FILHO; RICHETTI, 2014).

Currently, there are procedures in place to make it compulsory to have a doctor's prescription, but they don't cover all medicines. This is one of the reasons for the problem of high consumption, which leads to complications when abused, and if not used, we are faced with the waste caused by over-consumption of medicines. But it is not only self-medication that can lead to this problem, but also inappropriate prescribing, often caused by miscalculation of the time required for the drug to be effective in the body or even misdiagnosis (HOPPE; ARAÚJO, 2012).

Consumers are not encouraged to use medicines correctly, and when they don't cause individual accidents, we are faced with a huge amount of waste. And due to a lack of guidance and even alternatives, these expired or no longer used medicines are disposed of in inappropriate places. Incorrect disposal is one of the three causes of drug poisoning, along with autointoxication and accidental poisoning with children (TRIBESS JUNIOR; ZANCANARO, 2013).

Because they are not properly disposed of, medicines are thrown away in the garbage, sinks and toilets, which leads to soil and water contamination. Studies show that several of these substances appear to be persistent in the environment and are not completely removed in sewage treatment plants (MAIA; GIORDANO, 2012). These substances will compromise the health of the population, who may be subject to poisoning from ingesting this water contaminated by the drugs, whose toxic substances are able to resist the treatment process. And they may also be related to contact with soil or food from this environment.

There is no law obliging pharmacies to manage unused or expired medicines after they have been sold to the customer. According to Maia and Giordano (2012, p. 25)

> Given the lack of management and administration (public policies) in the disposal of liquid and solid medicines, the population's question arises as to where to dispose of these products and how to dispose of them correctly. If we obtain socio-environmental education, the support that society will have will help them as citizens and consequently help the environment.

Unfortunately, there is no appropriate place to dispose of medicines properly. Consumers are not made aware of this problem and so the lack of an alternative results in improper disposal, as there is no perception or knowledge of the harmful effects that these substances will cause when discarded in the environment. Irrational disposal not only compromises humanity's quality of life, but also causes an imbalance in the natural environment, compromising the fauna and flora present.

Despite the fact that the lack of an awareness program and publicity have increased the advertising and supply of medicines, and also because there is no specific procedure for the correct disposal of medicines, there is still a small part of the population that is aware of this issue.

One of the most discussed topics today is sustainable development. Concern about socio-environmental issues is taking on ever greater proportions. The subject is debated in the public and private sectors, as well as in civil society in general, because sustainability integrates the need for economic prosperity with social well-being and the protection and preservation of the environment. This concern makes us more discerning about consumption, as the irrational use of medicines will result in social and environmental problems (WESCHENFELDER, 2013).

Rational use guarantees a reduction in waste by consuming only what is necessary, without thinking about building so-called "home pharmacies". This change in attitude will reduce the amount of leftovers, which will result in less waste being generated, as well as being an act of solidarity when these medicines that are no longer used can benefit other people, who would no longer need to buy these medicines if they were handed over to collection points for redistribution.

The environment's capacity to regenerate is much smaller and slower than the production of waste, which is why the amount of waste generated is considered a serious environmental problem. Therefore, in order to overcome the unsustainable economic logic resulting from the Cartesian view, a new perception is needed that guides an environmental rationality (HOPPE; ARAÚJO, 2012).

In order for this awareness to become widespread, it is necessary to promote forms of awareness-raising so that knowledge about the problems of improper disposal of medicines can be increased. Building a harmonious relationship between humanity and nature will lead to citizens who are committed to environmental education.

There are many obstacles to carrying out this awareness-raising work, but that is why everyone's dedication and collaboration are important in order to reverse this situation, because according to Rodrigues (2009, p.80) "it is worth pointing out that the country's sanitary reality, with precarious infrastructure and a lack of sanitary landfills, is another factor that hinders the proper treatment of waste of a biological or chemical nature." There is a lack of incentive from public policies to make this practice a reality. It's not enough for citizens to be aware of this; public space must be made available and viable for the development of sustainable and solidarity-based activities.

Everyone has the right to a better quality of life, but collective responsibility is needed, emphasizing the consequences when practices are adopted properly. It is important that there is an emotional bond that will result in changes in attitudes, guaranteeing coexistence in a favorable environment. According to Hoppe and Araújo (2012), as long as man does not realize that he is part of the environment and that he is not above it, both he and nature will be increasingly harmed, but if there is such awareness, there will be a balance leading to a healthy and satisfactory integration.

Chapter 4

4 MATERIALS AND METHODS

For the development of this work, qualitative and quantitative research was carried out to obtain the desired data and apply contextualized methodology using questionnaires, textbook analysis and lectures and experiments that can relate the contents of organic chemistry and medicines in order to promote student interest.

First of all, visits were made to the Senador Levindo Coelho State School in the city of Ubá, Minas Gerais. This school was chosen because it has a laboratory where the experimental classes could be carried out, and also because the results were significant in the development of this work, which was carried out in eight 3rd year high school classes, since it is in this last stage of basic education that the students learn the content of organic fungi.

In order to find out how chemistry teachers develop the content of organic fungi, we used a structured questionnaire (Appendix I) with open and closed questions, which were applied to two teachers who teach in the 3rd year of regular secondary school, thus making it possible to find out what strategies they use to teach chemistry, whether they carry out experimental lessons or whether they use subjects that promote a relationship with the students' daily lives.

At the same time, structured questionnaires were administered to eight classes (Appendix II) totaling 205 students, with open and closed questions to understand their affinities and difficulties with the subject of chemistry, the recognition of substances and reactions that occur in everyday life and their previous knowledge of organic compounds, taking into account that the teachers had not yet introduced this content.

At the same time, we analyzed the organic chemistry content in the first two chapters of the textbook Química Cidada, volume 3 by Wildson Santos and Gerson Mol, which is the textbook adopted by the school and the support material used by the teachers to plan and carry out their lessons. The aim of this documentary analysis was to understand how the organic chemistry content is laid out and what links this content has with the subject of medicines.

Later, in order to promote a differentiated approach, periodic bibliographical reviews were carried out using various sources to prepare all the teaching material to be used, such as: content to be worked on, questionnaires, directed studies, selection of questions and scripts for lectures and experiments. The material was prepared by establishing the best conditions to be developed, relating medicines and functional groups with the aim of promoting interest in solving problems related to the students' daily lives through a contextualized approach with subjects pertinent to (CTSA).

After preparing the materials, the knowledge was applied through lectures on the subject of medicines, self-medication and their peculiarities. These lessons were carried out in the classrooms using Information and Communication Technology (ICT) and visual aids provided by the school. To apply the concepts of organic fungi, images were used of the structures of medicines that are present in the students' daily lives, such as paracetamol, aspirin, vitamin C and codeine. At this stage, subjects were used that promoted contextualization in the face of an investigative teaching methodology, with the insertion of problems related to the inappropriate use of medicines.

Experimental lessons were then carried out in the school's laboratory using its glassware. The reagents used were provided by the State University of Minas Gerais (UEMG) and the medicines were purchased from pharmacies. The practicals were investigative in nature and used recyclable packaging, in which the students identified which medicine was the correct one to use when faced with the proposed problem. The practicals were structured as follows: an introduction to the topic; a challenge involving solving the proposed problem; a procedure for developing the steps and post-practical questions for later analysis of the understanding of the content applied.

The following experimental lessons were developed as an investigative proposal:

1) Identification of phenol in paracetamol, a drug used to treat dengue fever (Appendix III);

2) Identification of antacids: sodium bicarbonate and aluminum hydroxide (Pepsamar) by reaction with hydrochloric acid (Appendix IV);

3) Similarities and differences between aldehydes and ketones in a proposal for identifying formaldehyde (Appendix V).

Finally, the data collected through the final questionnaire (Appendix VI) was analyzed in order to understand the evolution of learning performance by making a comparison before and after the different methodologies were implemented.

Chapter 5

5 RESULTS AND DISCUSSION

5.1 Teacher questionnaires

Initially, the professional and educational background of the two teachers was discussed. They both graduated from public institutions in the area in which they work and have been teaching chemistry at basic level for over ten years. One of the teachers only works at this school and the other is still vice-principal at another state school.

Analyzing the teachers' conceptions of the difficulty students have in learning chemistry, they believe that the reasons are related to a lack of interest and a gap in learning essential subjects such as mathematics. This makes it necessary for teachers to use methodologies that make it possible to broaden students' knowledge. Leite and Duarte (2014) state that the learning gap may be related to the lack of comparison between real and cultural subjects, creating a barrier to understanding scientific concepts.

When it came to using the textbook as a learning tool for applying scientific knowledge, the teachers reported using this material as their main working tool, and it is made available to all high school students. The textbook is essential to teachers' pedagogical practices, a fact illustrated by Fernandes and Porto (2012) who state that textbooks contain educational concepts that favor scientific knowledge, making them an important tool in the educational process. It is used by teachers to explain content, along with experimental lessons that are contextualized with the students' daily lives, which helps in the learning process.

The influence of the use of textbooks is also shown in the work of Bueno (2003), who states that it is of the utmost importance to articulate the content set out in textbooks with different teaching methodologies, correlating theory and practice with relevant themes and contributing to the formation of cognitive development, enabling the ability to think and understand everyday issues.

When analyzing the strategies the teachers use to arouse student interest, they point to practical lessons. However, both teachers said they did this infrequently, justifying it by a lack of time, as they considered the two classes a week to be insufficient to combine the exposition of content with experimental activities. For this reason, they leave it up to the undergraduates of the Institutional Teaching Initiation Scholarship Program (PIBID) to carry out monitoring and practical classes at different times. The presence of PIBID in schools has helped to apply different teaching methodologies, and is an important collaboration of the Commission for the Improvement of Higher Education Personnel (CAPES) for the development of pedagogical practices that favor the learning process with subjects pertinent to society. However, not everyone has access to these monitorships and so many students are deprived of this possibility of promoting knowledge mediated by experimental activities.

The results of the work by Firme and Amaral (2008) are similar to this one when they analyzed the teachers' conceptions of the incorporation of issues related to the CTSA in the development of theoretical and practical lessons, recognizing that the science-society relationship and experimental lessons have an influence on the construction of scientific knowledge.

Whenever possible, these experimental lessons, mediated by the teachers, use materials such as Styrofoam balls or modeling clay to assemble organic molecules. These lessons take place in the school laboratory so that the students can familiarize themselves with the environment, providing a differentiated lesson.

5.2 Student questionnaires

The students also took part in the research by answering a questionnaire, which was administered to eight 3rd grade classes, classified in the morning as 3°A to 3°F and in the afternoon as 3°G and 3°H, totaling 205 students in this initial phase. The questions were structured to obtain the following results:

1) Students' affinity with the subject of chemistry;

2) The reasons for difficulties in understanding the subject;

3) Addressing everyday situations such as identifying products and reactions that occur on a daily basis;

4) Identification of organic substances present in everyday life, taking into account that they had not yet studied this content, but with the aim of raising their prior knowledge.

Graph 1 shows that 20.5% of students answered YES, 21.5% answered NO and the majority of students, 58%, chose SOMETIMES. The most common response from those who like chemistry is related to interest and curiosity about the phenomena observed in a justification:

"I like studying chemistry because it arouses my curiosity and increases my understanding of certain phenomena. "

The majority of students opted for ONCE, which is quite interesting, considering that the methodology used is decisive in arousing students' interest in the content.

However, those who don't like chemistry usually justify it by their difficulty with formulas, calculations and the content.

The questionnaire also sought to explore the reasons behind the difficulties encountered by students in chemistry. Looking at Graph 1, most of the students, 71%, said that this difficulty was related to teaching using formulas, calculations, the lack of practical lessons and the relationship with everyday life, coinciding with the answers that lead students to dislike this subject. The answers provided show that some students highlighted methodologies that can contribute to learning, such as the need for practical lessons:

"Yes, sometimes you need more practical lessons to understand properly."

The results allow us to analyze that the students' difficulties may be related to the methodology in which the chemistry contents are applied, since the ways in which they are approached can contribute to low performance, especially when the theoretical concepts are presented as something that must be memorized, without making a relationship with aspects of everyday life, distancing themselves from the students' reality (OLIVEIRA, 2016).

The relationship between chemistry and everyday life is addressed in the third question. The data obtained is shown in Graph 1, where 72% of the students answered YES, giving examples of chemical reactions that occur around them on a daily basis, such as oxidation and combustion. This can be explained by the fact that they have more contact with these reaction processes, which allows them to observe the phenomena. However, 28% said NO. Of the results, only 3.7% of the students cited in their answers the use of medicines, which promote reactions in our bodies.

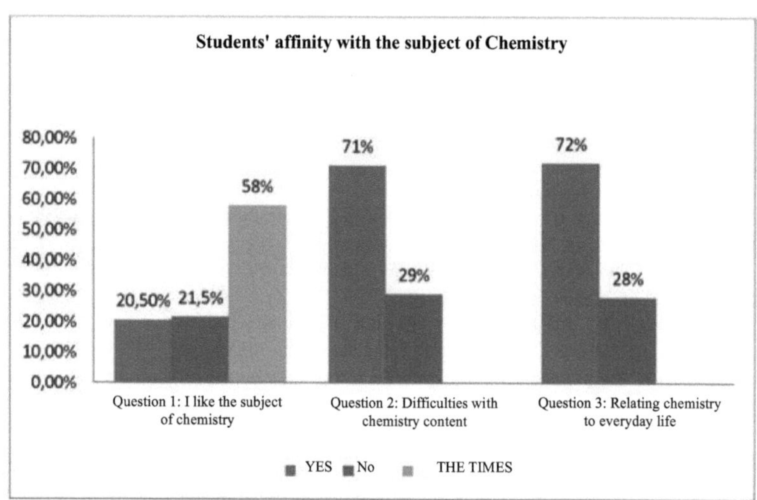

Graph 1- Analysis of the students' affinity with the subject of chemistry, relating their likes, difficulties and understanding of content related to everyday life

As a way of understanding the perception of identifying chemical products in everyday life, the students were asked to describe substances they lived with. It is important to note that cleaning products and cosmetics were the main products presented. They still associate chemical substances with something that has undergone an industrialized process. Medicines were represented by only 4.3% of the total number of students. Only one answered: "Everything around us is chemistry". This student understands the importance of chemistry, being present in everything around us. It is present in industrial processes, in economic and technological development, because all sectors use something of chemical origin as a raw material, since this science studies the structures of substances, their properties, and the transformations involved.

In order to study the students' prior knowledge of organic chemistry content, they were asked questions relating organic products present in everyday life. The analysis of the results, shown in Table 2, shows that the students do not have a formed concept of organic compounds or their relationship with substances present in everyday life, as seen when they consider inorganic compounds such as carbon dioxide, water, table salt and sodium bicarbonate to be organic compounds.

PRODUCTS CONSIDERED ORGANIC	PERCENTAGE
Carbon dioxide	**62,0%**
Oil	59,5%
Fat	58,0%
Water	**48,3%**
Table salt	**41,0%**

Sodium bicarbonate	41,0%
Alcohol	37,6%
Acetone	37,0%
Cooking oil	36,6%
Petrol	35,6%
Cooking gas	33,6%
Vinegar	31,7%
Butter	31,2%
Margarine	26,3%
Medicines	25,8%
Detergents	11,7%

Table 1 - Substances considered organic in the students' daily lives

This concern is expressed in the work of Quadros (2011), who states that studying chemistry has been a worrying responsibility given the results obtained by the students' conceptualization in defining and identifying the concepts belonging to this science. The lack of adequate tools prevents scientific knowledge from being inserted into students' lives.

5.3 Textbook analysis

The textbook should not be the only teaching resource used in class. The teacher's creativity in socializing the content, the use of this tool can favour the assimilation of knowledge. The textbook is an important didactic resource in the educational process when presented in a way that promotes the construction of chemistry concepts through proposals that arouse the students' interest (SILVA *et al,* 2013).

The book Química Cidada, used by the school and made available to all students, develops the content by bringing it closer to the students' everyday lives. Each chapter presents topics that develop these issues by presenting social themes, sustainable attitudes and experiments related to the content. Oliveira (2014) points out that chemistry teaching should relate social and personal aspects as a proposal to consolidate chemistry teaching.

In the development of the organic fungi content, presented in Chapter 2, this work begins with the insertion of "Food" as the theme in focus, emphasizing the economic importance of chemistry in the growth of the food industry. Marcondes (2008) reports that technological applications in the development of chemistry content contribute to scientific knowledge when linked to environmental, social, political and economic issues.

The topic on sustainable attitude presents informative questions on tips for healthier eating, instructing on care when obtaining, preparing and consuming food. Sustainability is an important subject that should be addressed in the development of chemistry content. The explanatory aspects that promote social and environmental responsibility can be fundamental tools for obtaining more comprehensive knowledge (DIAS, 2011).

After introducing each organic function, the book presents the application of these compounds, so that students can relate them to the reality they experience. According to Marcondes (2008), when the teacher uses facts that belong to the students' culture, contextualization becomes motivated by questions that allow for a more critical reading of the world.

This book makes it possible to implement proposals related to social and environmental issues, but with an emphasis on food-related subjects in the content of organic fungi. The topic of medicines was not chosen by the authors as a motivating theme for the content in question. The subjects involving medicines are only presented in the form of exercises. Richetti and Alves Filho (2014) point out that the influence of drug advertising establishes a theme that generates discussion in the classroom and can be associated with chemistry content.

Even though the subject of Medicines is not included, the organic chemistry content applied in the first two chapters of the book meets a contextualization proposal when it addresses current issues and makes a correlation with problems associated with inadequate nutrition.

5.4 Lectures

The lecture was the first step in introducing a different methodology, introducing concepts of organic chemistry through the theme of Medicines. Firstly, the difference between organic and inorganic compounds was presented, and the students were asked to identify substances such as methane, butane, ethyl methanoate and carbonic acid. They showed a lot of interest in the methodology used to apply the content using visual equipment which, combined with provocative dialogues, led to greater participation.

In the work of Malafaia *et al.* (2010), the use of differentiated didactic resources such as audiovisuals and computer tools is verified, as this methodology holds the students' attention. The use of technological strategies helps in the understanding of content, making them interdependent in educational processes in the face of the habits experienced. The application of the lecture is shown in Figure 1.

The organic chemistry content had already been presented in the classroom, and was added by the teachers after the previous questionnaire had been applied. In order to develop this methodology, a summary of organic fungi was presented, in which the concepts were defined by presenting the structures of medicines that are present in the students' daily lives. This application is of great relevance, and although this term contextualization is much discussed, few teachers teach their classes addressing issues that can contribute to students' learning (BRAIBANTE, 2010).

The students helped to identify the following organic fungi: alcohol, ether and amine in codeine; alcohol, enol and ester in vitamin C; phenol and amide in paracetamol and carboxylic acid and ester in aspirin. It was observed that the application of this methodology led to greater interest in relating the content to issues in the students' lives, enabling a collective discussion of the topics covered. This result corroborates the work of PAZINATO *et al.*, (2012), who found that, as it is a subject of social debate, the use of the theme of medicines motivated learning, since numerous organic fungi are present in their molecules, making this subject more attractive for the development of this content.

5.5 Experimental classes

Experimental lessons are moments in which the content introduced is problematized, stimulating a process of investigation. Through experiments, it is possible to promote students' curiosity and also allow knowledge to become a reality in the learning process.

In this research, the experimental lessons were developed to stimulate the students' ability to problematize and investigate. The first experiment (Appendix III) is relevant because it deals with current affairs. As an investigative proposal, two painkillers/antipyretics were presented: aspirin and paracetamol. The students recognized the correct medicine to use to treat the symptoms of dengue by identifying an organic function present only in paracetamol. To carry out this practice, we used the two medicines, methanol and ferric chloride shown in Figure 2.

Figure 2 - Materials used to apply the first practical lesson to identify the organic substance phenol

The introduction dealt with problems related to dengue fever and the adverse effects of using inappropriate medication. The development of this practice is shown in Figure 3.

Figure 3 - Illustration of the students' participation in practical lesson 1

In a challenging proposal, the students identified the two drugs by reacting them with the ferric chloride shown below:

$$3\ Ar\!-\!OH_{(s)} \ + \ FeCl_{3(aq)} \longrightarrow \ Fe(OAr)_{3(aq)} \ + \ 3\,HCl_{(aq)}$$

In view of the results observed, the complexation reaction occurs due to the presence of the phenolic hydroxyl present only in paracetamol, forming a colored compound as shown in Figure 4, where paracetamol is represented by the greenish colored test tube and aspirin, because it does not have the phenol functional group, does not react, observed by the yellow colored test tube which corresponds to the color of ferric chloride.

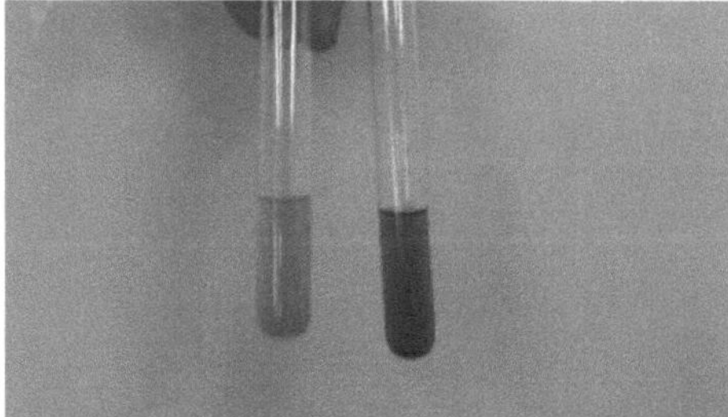

Figure 4 - Reaction to identify phenol in paracetamol forming a greenish compound

The compound shown in the tube on the right is paracetamol and the one on the left is aspirin. This curiosity in determining which would be the correct medicine to use to treat the symptoms of dengue fever led to greater learning in identifying the fungi present in these medicines. This positive result can be seen in the answers obtained in the questionnaires after applying this practice. Investigative experiments promote the construction of knowledge where the teacher is only the mediator, in line with the studies by Wilsek, (2012) where investigative activities provide greater curiosity in students which favors a collective dialogue in solving problematic situations.

The second experiment (Appendix VI) dealt with two antacids, sodium bicarbonate and aluminum hydroxide respectively. The materials used in this experiment are shown in Figure 5.

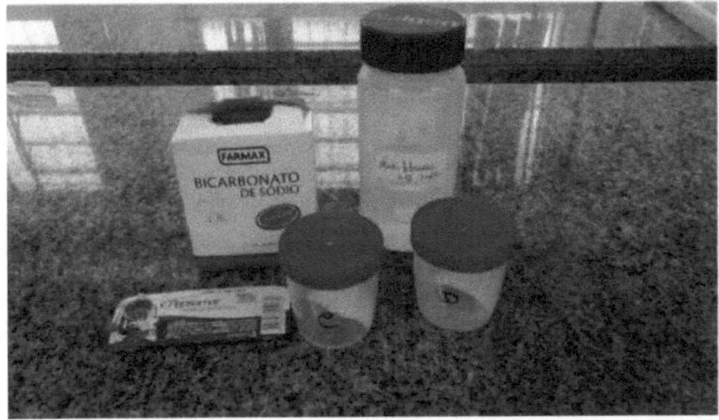

Figure 5 - Materials used to apply the second practical lesson in the identification of two antacids

The students identified these drugs through a neutralization reaction with hydrochloric acid, and it was possible to observe the release of carbon dioxide only in the sodium bicarbonate reaction.

During the practical, subjects related to self-medication and the health risks of overusing these antacids were also discussed, with an emphasis on the problems related to sodium bicarbonate in hypertensive people. Figure 6 illustrates this practice in one of the classes.

Figure 6 - Illustration of the students' participation in practical lesson 2

The approach to problems related to the use of sodium bicarbonate in hypertensive people aroused the curiosity of the students in carrying out the practice, since these drugs can have adverse effects according to each person's body. Here are the reactions that occur when these drugs are used:

$$Al(OH)_{3(s)} + 3HCl_{(aq)} \rightarrow AlCl_{3(s)} + 3H_2O_{(l)}$$
$$NaHCO_{3(s)} + HCl_{(aq)} \rightarrow NaCl_{(s)} + H_2O_{(l)} + CO_{2(g)}$$

The arrangement of the reactions of the two antacids: sodium bicarbonate and aluminum hydroxide with hydrochloric acid favored a preliminary observation so that the students could analyze the products formed and better understand the damage caused by the inappropriate use of these medicines, since when sodium bicarbonate reacts with hydrochloric acid present in the stomach, sodium chloride (NaCl) is formed.

This practice was interesting because of the health-related issues, as many students said they used these medicines, but were unaware of the related problems. Thus, according to Alves Filho and Richetti (2014), it is possible to associate social problems to arouse students' interest when related to scientific knowledge, enabling an approach through the creation of problem situations (OLIVEIRA, 2012).

The third experiment (Appendix V) *looked* at the organic compounds aldehydes and ketones, their main differences and similarities and the applicability of these compounds in our daily lives. To carry out this experiment, two substances familiar to the students were used: formaldehyde and acetone. Formaldehyde is used as a disinfectant, in the preservation of anatomical handles, in the manufacture of plastics and compounds such as medicines and explosives. Acetone is a synthetic intermediate widely used as a solvent and remover.

To prepare the Tollens reagent, used to identify the aldehyde functional group, the following reagents were used: sodium hydroxide (NaOH), silver nitrate ($AgNO_3$) and ammonium hydroxide (NH_4OH). These substances were placed in coded bottles where the students identified the formaldehyde by reacting it with Tollens' reagent to form the silver mirror, shown in Figure 7.

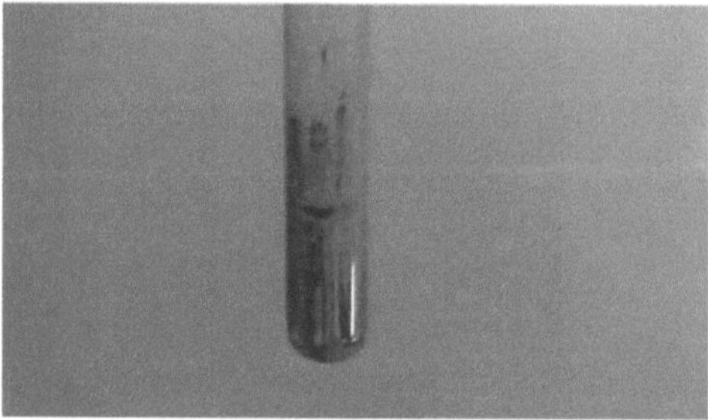

Figure 7- Formation of the silver mirror in the reaction of formaldehyde with Tollens' reagent

The challenge of identifying organic fungi present in everyday substances gave the students an interest in the subject. The choice of these reagents encouraged the students to identify them because they recognized their applications. Below are the reactions for forming Tollens' regent (Reaction 1 and 2).

Reaction 1)
$$2\ AgNO_{3(aq)} + 2\ NaOH_{(aq)} \rightarrow Ag_2O_{(s)} + H_2O_{(l)} + 2\ NaNO_{3(aq)}$$

Reaction 2)
$$AgO_{2\ (s)} + 4\ NH_{3(aq)} + H_2O_{(l)} \rightarrow 2\ Ag(NH_3)_2OH_{(aq)} \text{ (Reagente de Tollens)}$$

Tollens' reagent is a complex in a basic medium. This reagent oxidizes the aldehyde to a carboxylic acid and at the same time reduces silver with the formation of metallic silver, which is deposited on the wall of the test tube, forming the mirror observed in the experiment. The reaction to form the silver precipitate in the identification of the aldehyde is shown in Reaction 3 below:

Reaction 3)

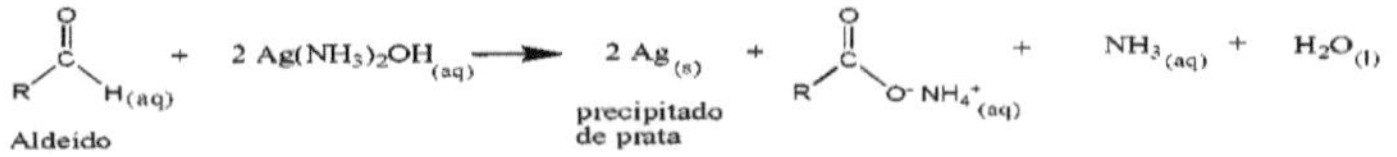

Aldeído: $R-CHO_{(aq)}$ + $2\ Ag(NH_3)_2OH_{(aq)} \longrightarrow 2\ Ag_{(s)}$ (precipitado de prata) + $R-COO^-\,NH_4^+{}_{(aq)}$ + $NH_{3(aq)}$ + $H_2O_{(l)}$

Silver is a cation that has oxidation-reduction properties, changing the color of solutions in the presence of analytical reagents. With Tollens' reagent, silver changes from Ag^+ to Ag^0, forming a black precipitate capable of mirroring. This methodology is the basis for the manufacture of mirrors (DEMIATE, 2009).

The contextualization with objects present in the students' daily lives made it possible to gain a greater understanding, as seen in the answers obtained from the post-practice questions and also in the visual aspect when they managed to obtain the desired product. The use of methodological resources through experimentation is fundamental for scientific learning, as visualization arouses interest and questions about the content. The dissemination of information needs to be approached experimentally in a clear way, bringing the subject of chemistry closer to students' everyday lives (MALAFAIA, 2010; REGINALDO, 2012).

5.6 Final questionnaire

Organic substances are defined as substances that have carbon atoms. These atoms bind together to form carbon chains that carry out numerous reactions. Many molecules have carbon due to the location of this element in the periodic table, which makes it possible to share electrons and consequently form various stable compounds. Most of the substances found in nature are organic, such as those used for food, medicine and clothing. They are very important, but not limited to this, as chemists with their knowledge are able to synthesize organic compounds from even substances never found in nature (BRUICE, 2006).

As a way of understanding the assimilation of content after applying the contextualized methodology, this questionnaire aimed to verify the students' understanding of the concept of organic function, as well as the contextualization with the proposed theme. Graph 2 shows that 92% identified that an organic compound has carbon atoms linked in a chain and/or carbon atoms linked directly to hydrogen.

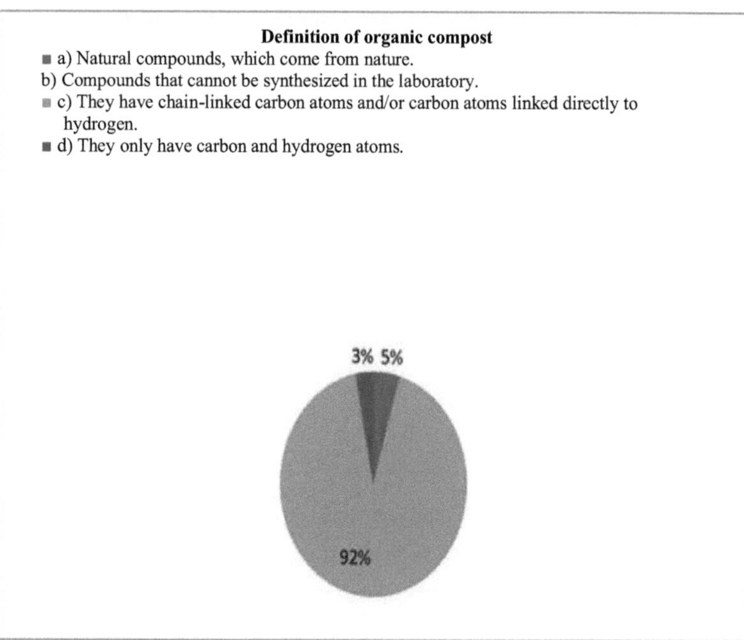

Graph 2 - Representing the students' answers when defining an organic compound

After using a methodology aimed at tackling issues relevant to society from a CTSA perspective, the analysis of the second question in the final questionnaire sought to understand the students' ability to distinguish between organic and inorganic compounds present in their daily lives, such as: medicines, cooking gas, alcohol, vinegar, butter, caustic soda, water and table salt. Graph 3 shows that 85% of the students were able to identify the correct answer, representing medicines, cooking gas and alcohol as organic compounds.

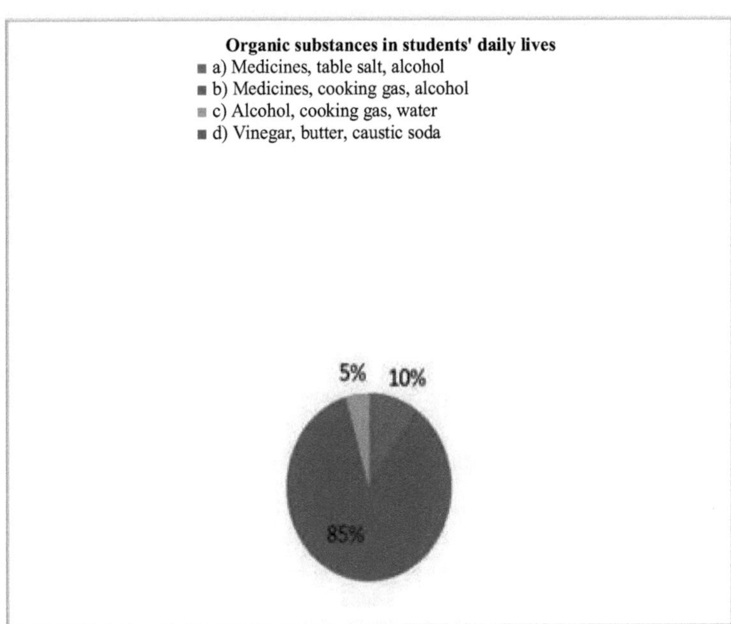

Graph 3 - Identification of substances present in students' daily lives

This quiz also sought to understand the students' ability to recognize the molecular formulas of organic compounds and differentiate them from inorganic compounds, which was analyzed by question three and observed by the results in Graph 4, where 93% of the students identified the correct answer which included the three organic compounds: C_3H_6O (acetone), CH_4 (methane) and CH_3COOH (acetic acid).

The students were generally able to identify and relate the substances around them to their respective molecular formulas. This contextualization brings learning closer to reality, without the need to memorize molecules and reactions (BRUICE, 2006).

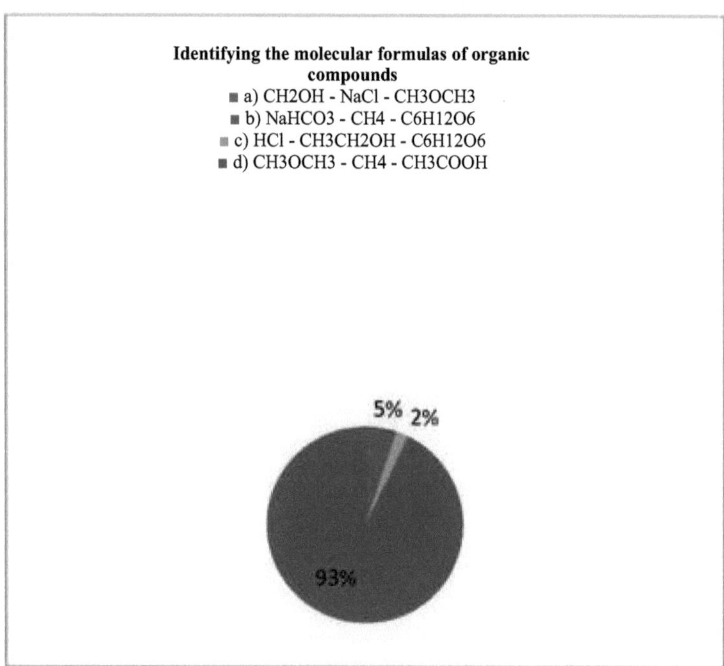

Graph 4 - Understanding of students' ability to identify molecular formulas of organic compounds

The students' knowledge of identifying organic fungi present in everyday substances such as alcohol, vinegar, cooking gas and acetone was addressed in the fourth question, where 97% of the students were able to relate these substances to their respective organic fungi, as can be seen in the results shown in Graph 5. Chemistry teaching does not yet provide sufficient conditions for students to understand the concepts and be able to relate them to everyday life. In order to do this, lessons need to be taught in a way that seeks out points of articulation that can relate to the student's current knowledge.

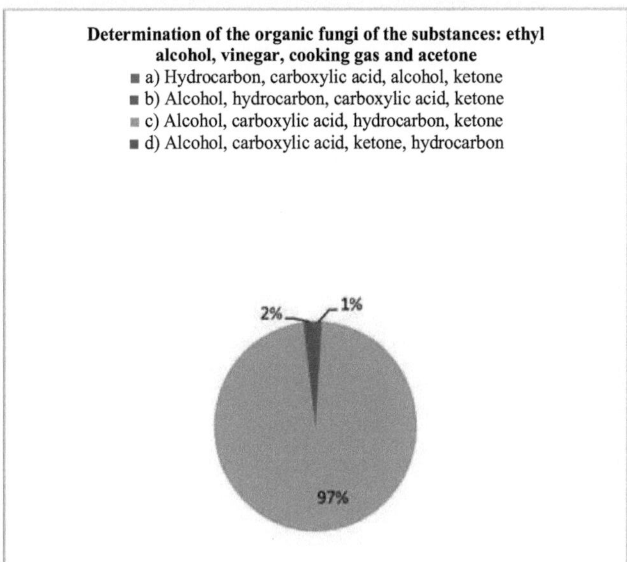

Graph 5 - Relationship between organic fungi and substances found in students' everyday lives

Relating the content covered to the subject of medicines, questions 5 to 8 were aimed at identifying the organic fungi present in four medicines: paracetamol, aspirin, codeine and vitamin C. The students' performance in the questions is shown in Graph 6 on page 42.

The first structure to be presented was paracetamol, as shown in Figure 8, where the organic fungi amide and phenol are present. Looking at Graph 6, it can be seen that 82% got this question right, making it possible to identify that, like the work of Eichler and Gorri (2013), the presentation of contextualized figures makes it possible to develop broader learning, leading students to understand the microscopic world, not just macroscopic properties and their changes.

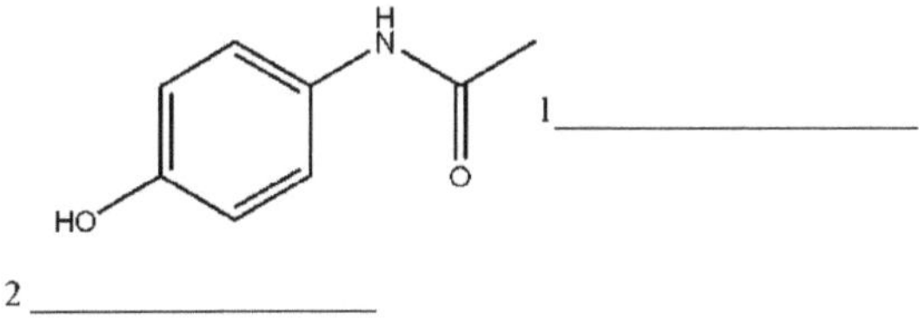

Figure 5- Structure of paracetamol

The molecular formula of aspirin is shown in Figure 9, which contains the organic compounds carboxylic acid and ester. Graph 6 shows that 89% got this question right. From the errors obtained, it can be seen that some students confused ester with ketone due to the presence of C=O.

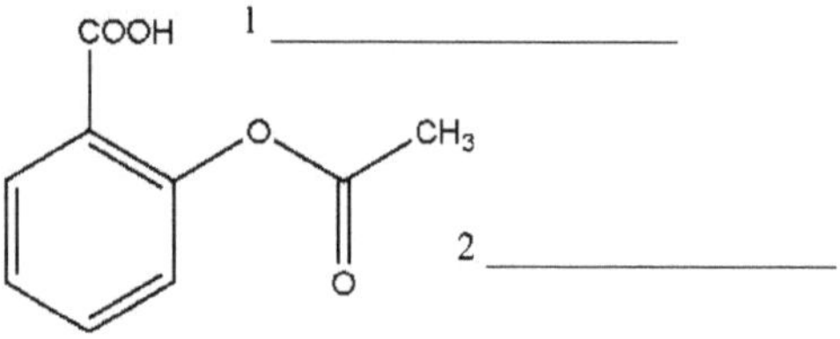

Figure 9- Structure of aspirin

With regard to the structure of codeine, Figure 10 shows that a small number of students, 15%, had difficulty differentiating between amine and amide, ether and ester and alcohol and enol. However, most of the students, 85%, correctly identified these fungi as: ether, ether, alcohol and amine respectively.

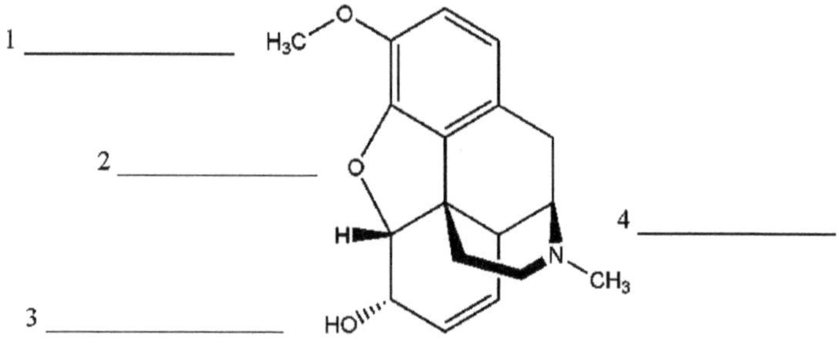

Figure 10 - Structure of codeine

For the results obtained in identifying the organic fungi present in Vitamin C, shown in Figure 11, it can be seen that the majority of students, 81%, got this question right, with the correct alternative being the one that presents the organic fungi: alcohol, alcohol, enol, enol and ester.

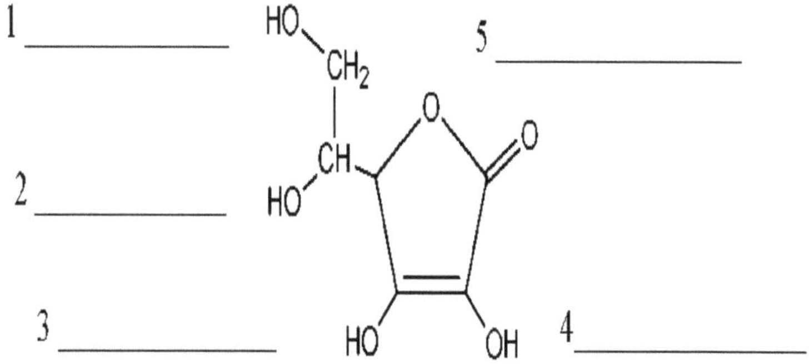

Figure 11 - Vitamin C structure

Some students found it difficult to identify the ester function with ether ketone. Other students, 9%, identified the organic compounds represented by three and four as alcohol, whereas the correct figure is enol, because the hydroxyl is attached to an unsaturated carbon. The majority of students, 81%, got this question right, which presents the organic fungi: alcohol, alcohol, enol, enol and ester.

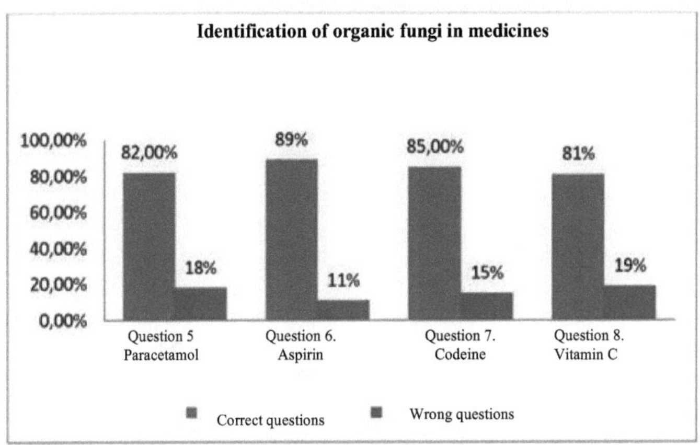

Graph 6 - Answers obtained when identifying the organic functions present in four medicines

In order to understand the students' assimilation of the issues raised in the experimental lesson, question nine addressed issues related to two antacids: sodium bicarbonate and aluminum hydroxide. They mentioned which medicine would be more suitable for a hypertensive person based on the products generated by these reactions.

Graph 7 shows the students' understanding of the contextualization of this content with problems related to self-medication, when 97% answered sodium hydroxide as the most appropriate medicine to use. This positive result is justified by visualizing the reactions that occur between medicines and hydrochloric acid, also covered in the second experimental lesson, in which sodium chloride (NaCl) is formed in the sodium bicarbonate reaction.

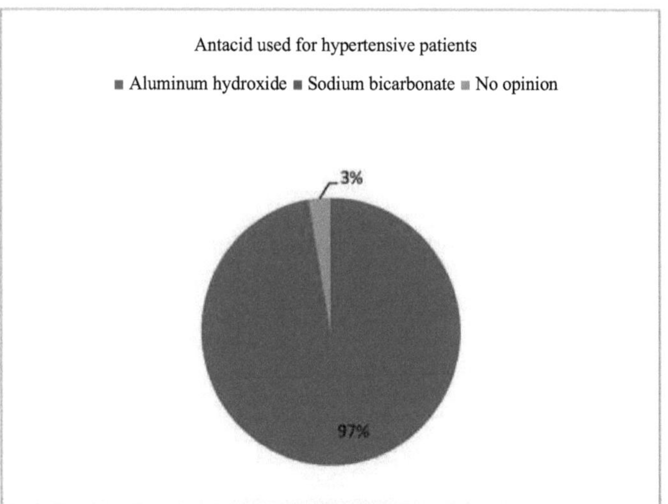

Graph 7 - Identifying the right medication for hypertensive people by observing reactions with hydrochloric acid

In order to investigate the absorption of content related to the similarities and differences between the organic fungi aldehydes and ketones, question ten presented the structural formulas of two substances from the students' everyday lives: formaldehyde and ketone, shown in Figure 12.

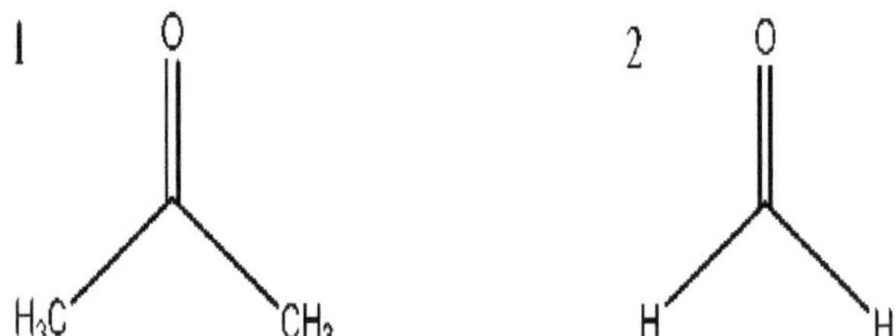

Figure 12 - Structural formula of acetone and formaldehyde

Both structures have the carbonyl functional group. Structure 1 is propanone, also called acetone, which belongs to the ketone organic function, and structure 2 is formaldehyde, whose organic function is aldehyde. Most of the students, 89%, were able to identify the incorrect answer in which structure 1 is not formaldehyde, as shown in graph 8.

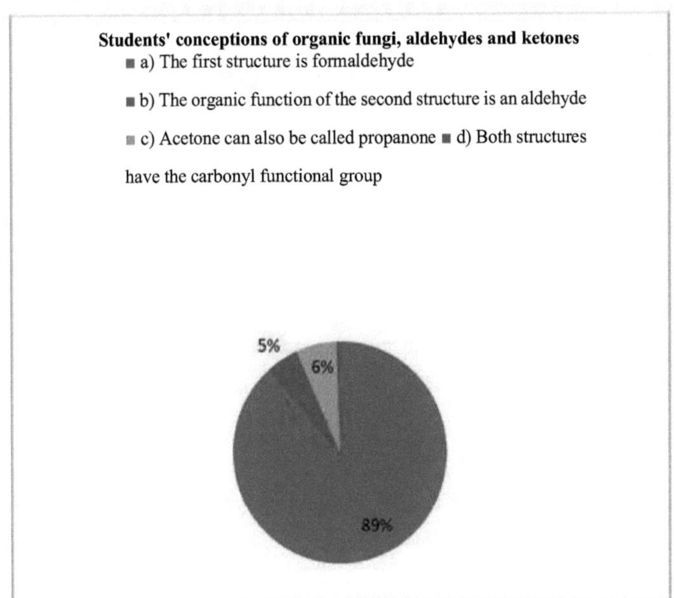

The use of the substances formaldehyde and acetone made it possible to learn more about the organic functions aldehydes and ketones. The result obtained in this question shows that the assimilation of chemistry content is related to understanding the phenomena that occur in the students' daily lives (RAUPP, PINO, 2014).

In addition to contextualizing with substances from the students' everyday lives, research should be promoted. The expectation of which product would form the silver mirror, addressed in the third experimental lesson, instigated curiosity in understanding the similarities and differences between the two functional groups, obtaining a positive result.

CONCLUSION

The reasons for the obstacles encountered by students in learning chemistry are linked to the way in which the content is taught. The students' conception of this content is based on a methodology with few experimental and contextualized lessons, which leads to difficulties in understanding. Due to the lack of time, the content is applied with the insertion of formulas, calculations and memorization. However, both teachers consider it essential to introduce expository and experimental lessons and to use current affairs as a mediating theme for teaching chemistry.

In order to be able to include differentiated resources, the school's pedagogical contribution is necessary to make the necessary timetables available for the development of this subject.

The textbook used by the teachers meets the didactic objective of contextualizing organic chemistry content with current affairs. However, the proposed theme is not observed as a mediator in the approach to this content. This material is a very important tool in the construction of knowledge, but it should be used with other methodologies.

The use of the theme Medicines in the application of expository and experimental lessons can be used to avoid the gap and fragmentation of content, as well as enabling students to think critically when solving proposed situations. In the development of organic fungi content, it can be applied in order to understand their structures and identify related problems.

It can be concluded that in order to achieve meaningful learning that enables the formation of critical citizens, different methodologies are needed that are contextualized with subjects from the students' daily lives.

The contribution of this work to the field of research in the area of chemistry teaching is the inclusion of methodologies that favor more critical learning with subjects of current relevance. As well as promoting knowledge of the content of organic fungi, the approach to the subject of Medicines can also be used to raise awareness of the impact on health of excessive consumption of medicines and their improper disposal.

REFERENCE

ALBA, J.; SALGADO, T. D. M.; DEL PINO, J. C. D. Case studies: a proposal for approaching organic chemistry fungi in high school. **Revista Brasileira C&T**, Porto Alegre, v. 6, n° 2, p. 76 - 96, mai./ago. 2013.

BERNARDES, P. O.; SILVEIRA, H. E. Conceptions of chemistry: an analysis of figures produced by basic education students. In: XV Encontro Nacional de Ensino de Química (XV ENEQ) - Brasília, DF, Brazil, 2010.

BRAIBANTE, H.T.S.; BRAIBANTE, M.E.F.; TREVISAN, M.C. and PAZINATO, M.S. Overhead projector as a chemistry laboratory bench. Santa Maria: Palotti, 2010.

BRAZIL. National Health Surveillance Agency. **What we should know about medicines.** Trecho 5, Área especial 57, Brasília, DF, 2010.

BRUICE, P. Y. **Química Orgánica**, vol. 1 ed. 4, 2006.

BUENO, L.; MOREIRA, K. C.; SOARES, M.; DANTAS D. J.; WIEZZEL, A. C. S.; TEIXEIRA, M. F. S. Teaching chemistry through experimental activities: the reality of teaching in schools. **Sao Paulo.** Universidade Estadual Paulista "Júlio de Mesquita Filho" Faculdade de Ciencias e Tecnologia, Presidente Prudente, 2007.

CARNIEL, V. L.; FARIAS, C.; MISTURA, C. M.; BOTH, J.; PELLIZZARI, R. R.;

CUNICO, N. S.; BEDIN, T. M. Recognizing the functional groups of Organic Chemistry through the contextualization of the study of medicines. In: CURRICULAR MOVEMENTS IN CHEMICAL EDUCATION, 33, 2012, Passo Fundo: UNIJUÍ, 2012.

DEMIATE, I. M.; WOSIACKI, G.; CZELUSNIAK, C.; NOGUEIRA, A. Determination of reducing and total sugars in foods. Comparison between colorimetric and titrimetric methods. **Exact and Earth Sciences, Agricultural Sciences and Engineering**, v. 8, n. 1, 2009.

DE OLIVEIRA, J. R. S. Contributions and approaches of the experimental activities in the science teaching: Gathering elements for the educational practice. **Acta Scientiae**, v. 12, n. 1, p. 139-153, 2012.

DE QUADROS, A. L.; DA SILVA, D. C.; DE ANDRADE F. P.; ALEME, H. G.; SILVA, G. F. Teaching and learning Chemistry: the perception of high school teachers. **Educar em Revista**, n. 40, 2011.

DIAS, R. Gestao ambiental: responsabilidade social e sustentabilidade. In: **Gestao ambiental: responsabilidade social e sustentabilidade**. Atlas, 2011.

STREAK, D. R.; REDIN, Euclides; ZITKOSK , J. J. **Dicionário de Paulo Freire**. 2. ed. Belo Horizonte: Autentica Editora, 2010.

DO NASCIMENTO FIRME, R.; DO AMARAL, E. M. R. Chemistry teachers' conceptions of science, technology, society and their interrelationships: a preliminary study for the development of CTS approaches in the classroom. **Ciencia & Educado**, v. 14, n. 2, p. 251-269, 2008.

DOS SANTOS, W. L. P. Contextualization in science teaching through CTS themes from a critical perspective. **Ciencia & Ensino**, v. 1, 2008.

FARY, B. A.; SOUZA, E.; SOATO, A. M. L.; TONIN, L. T. D.; BARON, A. M. Chemistry of medicines: teaching organic functions combined with social awareness. In: SIMPÓSIO NACIONAL DE ENSINO DE CIÉNCIAS E TECNOLOGIA, 3.,Sep. 2012, Ponta Grossa, 2012. Ponte Grossa: SINECT, 2012.

FERNANDES, M. A. M.; PORTO, P.A. Investigating the presence of the history of science in General Chemistry textbooks for higher education. **Química Nova**, v. 35, n. 2, p. 420429, 2012.

FERNANDES, I. M. B. **A perspectiva CTSA nos manuais escolares de ciencias da natureza do 2° CEB**. 2011. Doctoral thesis. Polytechnic Institute of Braganga, School of Education.

FERREIRA, L.H.; HARTWIG, D.R. e OLIVEIRA, R.C. Ensino Experimental de Química: Uma abordagem investigativa contextualizada. Química Nova na Escola, v.32, n.2, p.101106, 2010.

FERRAES, A. M. B.; LANGER, T. E. C. **"Rational Use of Medicines Project in Cornélio Procópio/2008 - preventing poisoning and promoting correct disposal".** Cornélio Procópio, 2011.

FREIRE, Paulo. **Education as the practice of freedom**. Editora Paz e Terra, 2014.

GIBIN, G. B.; FERREIRA, L. H. Avaliação dos estudantes sobre o uso de imagens como recurso auxiliar no ensino de conceitos químicos. **Química Nova na Escola**, v. 35, p. 19-26, 2012.

GORRI, A. P.; EICHLER, M. L. On the Language of Organic Chemistry: Acids, Bases and their Signs. **Encontró de Debates sobre o Ensino de Química**, v. 1, n. 01, 2013.

GUIMARAES, C. C. Experimentation in chemistry teaching: paths and detours towards meaningful learning. **Química Nova na Escola**, v. 31, n. 3, p. 198-202, 2009.

LIMA, J. O. G.; BARBOSA, L. K. A. Chemistry teaching in the conception of elementary school students: some reflections. **Exatas Online**, vol. 6, n.1, p. 33-48, 2015.

LEITE, R.; DUARTE, M. C. Teachers' perceptions of the concept of analogy and its use in the teaching and learning of Physics and Chemistry. **Analogies, Readings and Models in Science Teaching: the classroom under study. Escrituras**, v. 6, p. 45-49, 2014.

HOPPE, T. R. G.; ARAÚJO, L. E. B. Environmental Monographs, REMOA/ UFSM. Environmental contamination due to improper disposal of expired or unused medicines. **Monografías Ambientais**, v. 6, n° 6, p. 2236-1308, mar. 2012.

MAIA, M.; GIORDANO, F. Study of the current situation of awareness of the population of santos regarding the disposal of medicines. **Revista Ceciliana**, v. 4, n° 1, p. 24-28, jun. 2012.

MALAFAIA, G.; BÁRBARA, V. F.; RODRIGUES, A. S. L. Analysis of students' conceptions and opinions on the teaching of Biology. **Revista Eletronica de Educado**, p. 165182, 2010.

MARCONDES, M. E. R. Proposigoes metodológicas para o ensino de Química: oficinas temáticas para a aprendizagem da ciencia e o desenvolvimento da cidadania. **In Extension**, v. 7, n. 1, 2008.

MEDEIROS, A. S.; MORAIS, A. E. R.; LIMA, S. L. C.; REINALDO, S. M. A. S.; FERNANDES, P. R. N. Importance of practical classes in chemistry teaching. In: IX IFRN Scientific Initiation Congress. 2013.

MESQUITA, E. C.;, OLIVEIRA, A. I. D.;, ABADIA, G. J. S.; CARVALHO, C. V. M. Experimentation in chemistry teaching: an approach based on the identification of learning difficulties. In: IV State Congress of Scientific Initiation of IF Goiano, 2015.

OLIVEIRA, A. I. D.; MESQUITA, E. C.; SILVA, L. A. S.; MIRANDA, C. V. The use of experimentation in the early grades of high school to approach chemical content. **Ciclo Revista**, v. 1, n. 2, 2016.

OLIVEIRA, B. P. **Evaluation of Chemistry Textbooks and the Conceptions of High School and College Students on the Content of Spontaneous Phenomena**. Doctoral thesis. 2014.

PAZINATO, M. S.; BRAIBANTE, H. T. S.; BRAIBANTE, M. E. F.; TREVISAN, M. C.; SILVA G. S. A differentiated approach to teaching organic functions through the theme of medicines. **Revista Química Nova na Escola**, v. 34, n° 1, p. 21-25, feb. 2012.

RAUPP, D.; SERRANO, A.; MARTINS, T. L.C. The evolution of computational chemistry and its contribution to chemistry education. **Revista Liberato, Novo Hamburgo**, v. 9, n. 12, p. 13-22, 2008.

RAUPP, D. T.; DEL PINO, J. C. Stereochemistry in Higher Education: historicity and contextualization in Organic Chemistry textbooks. **Acta Scientiae**, v. 17, n. 1, 2014.

REBELLO, G. A. F. ARGYROS, M. M.; LEITE, W. L. L.; SANTOS, M. M.; BARROS, J. C.; DOS SANTOS, P. M. L.; DA SILVA; J. F. M. Nanotechnology, a theme for high school using the CTSA approach. **Química nova na escola**, v. 34, n. 01, p. 3-9, 2012.

REGINALDO, C.C.; SHEID, N. J.; GÜLLICH, R. I. C. Science Teaching and Experimentation. **Seminário de pesquisa em educaçao da regiao sul**, v. 9, p. 1-13, 2012.

RIBEIRO, T. V.; GENOVESE, L. G. R.; COLHERINHAS, G. Teaching by Research in High School: Discussing CTSA issues in scientific literacy. Technology. **VIII ENPEC-National Meeting of Research in Science Education. Minutes... State University of Campinas. Campinas, SP**, 2011.

RICHETTI, G. P.; ALVES FILHO, J. P. Self-medication in Chemistry Teaching: an interdisciplinary proposal for High School. **Revista Educación Química**, v. 25, n° 1, p. 203-209, abr. 2014.

RICHETTI, G. P.; ALVES FILHO, J. P. A Automedicaçao: um tema social para o Ensino de Química na perspectiva da Alfabetizacao Científica e Tecnológica. **Revista de Educaçao em Ciência e Tecnologia,** v. 2, n° 1, p. 85-108, mar. 2009.

RODRIGUES, C.R. B. **Legal and environmental aspects of medicine waste disposal.** 2009. 110f. Dissertation (Master's in Production Engineering) - Federal Technological University of Paraná, Ponte Grossa.

SAVIANI, D. **Education from common sense to philosophical consciousness**. Autores Associados, 2007.

SILVA, A. M. Proposal to make chemistry teaching more attractive. **Revista de**, 2011.

SILVA, G. S.; BRABANTE, M. E. F.; PAZINATO, M. S. The visual resources used in the approach to atomic models: an analysis in chemistry textbooks. **Revista Brasileira de Pesquisa em Educagao em Ciencias**, v. 13, n. 2, p. 159-182, 2013.

SILVA, R. T.; CURSINO, C. T.; AIRES, J. A.; GUIMARÁES, O. M. Contextualization and Experimentation An Analysis of Articles Published in the "Experimentation in Chemistry Teaching" Section of the journal Química Nova na Escola 2000-2008. **Ensaio Pesquisa em Educagao em Ciencias**, v. 11, n. 2, p. 245-261, 2009.

SILVA, V.S; KLUBER, T. E. Mathematical modeling in the early years of elementary school: an imperative investigation. Revista Eletrónica de Educagao, v. 6, n. 2, p. 228249, 2012.

SUTIL, N.; BORTOLETTO, A.; CARVALHO, W.; CARVALHO, L. M. O. CTS and CTSA in national journals in science/physics teaching (2000-2007): epistemological and sociological aspects. **XI Physics Teaching Research Meeting. Curitiba**, 2008.

TRIBESS JUNIOR, A.; ZANCANARO, V. Disposal of household medicines and environmental impact: Awareness of the population in the municipality of cagador/sc1. **Revista Extensao em foco**, v. 1, n° 1, p. 54-58, dec./ nov. 2013.

VYGOTSKY, L.S. **The social formation of the mind. Sao Paulo**: Martins Fontes, 1989. Available at: <http://egov.ufsc.br/portal/sites/default/files/vygotsky-a-formac3a7c3a3o-social-da-mente.pdf>. Accessed on: November 20, 2016.

WARTHA, E. J.; SILVA, E.; BEJARANO, N. R. R. Everyday life and contextualization in chemistry teaching. **Química Nova na Escola**, v. 35, n. 2, p. 84-91, 2013.

WESCHENFELDER, F. C. **The reverse logistics of medicines and their impact on sustainability:** A case study in the Dimed-Panvel group. 2013. 93f. Monograph (Bachelor in Administration) - Federal University of Rio Grande do Sul, Porto Alegre.

WILSEK, M. A. G.; TOSIN, J. A. P. Teaching and Learning Science in Elementary School with Investigative Activities through Problem Solving. **Estado do Paraná**, v. 3,

n. 5, p. 1686-8, 2012.
ZUIN, V. G.; FREITAS, D. The use of controversial themes: a case study in the training of undergraduates in a CTSA approach. **Ciencia & Ensino (ISSN 1980-8631)**, v. 1, n. 2, 2008.

APPENDIX

APPENDIX I: Teacher questionnaire

Survey Questionnaire

Teacher's name: ____

1. **Training data:**

a) What is your degree? _______________________________________

b)   Name of training institution:

____

c) Year completed: ____

2. **How long have you been an elementary school chemistry teacher?**

____

3. **Do you work in another educational institution or do you have more than one position in your school?**

( ) No ( ) Yes

If **yes**, please specify which institution and/or which position:

____

____

4. **Chemistry is considered a difficult subject to understand, do you notice these difficulties in students' learning?**

a) ( ) Yes. In your opinion, what are the reasons for this difficulty?

____
____
____

b) ( ) No. Justify

____
____

5. **Do you use textbooks as a learning tool?**

( ) No.
( ) Yes. Specify which book(s) you use, indicating author and volume:

____
____

a) Is the textbook in question available to all students?

( ) Yes (   ) No

6. **What strategies do you use in your classes to arouse students' interest?**

____

____

____

7. **Do you use situations from the students' everyday lives to teach chemistry?**

    ( ) Yes   ( ) No

    If the statement is **positive,** answer questions a, b and c.

    a) How do you work with these situations?

____

____

    b) Is it noticeable that the students are more interested in the lessons?
        ( ) Yes, ( ) No.
    c) Does learning become more meaningful? Justify your answer.
        ( ) No.
        ( ) Yes. ____

____

____

8. **Do you use any experimental classes?**
    ( ) No. Justify

____

____

    ( ) Yes
    If the statement is positive, answer questions a, b, c and d.
    a) What materials are used as support and why? (Alternative materials and/or
        commercial reagents).

____

____

    b) Does the school have these materials? ( ) Yes, ( ) No.
    c) How often are these experiments carried out and how long do they take?

____

____

    d) Are these experimental lessons held in classrooms and/or laboratories? Explain
        why you prefer the space.

____

____

9. **In your opinion, what are the contributions to learning when methodologies are used that promote relationships with the students' daily lives?**

____

____

____

10.   And in relation to the experimental classes, what contributions do you see in the student's   learning?

____

____

____

____

____

8.1 APPENDIX II: Students' preliminary questionnaire

Survey Questionnaire

Student's name: ____

Address:____

____

1. Do you like studying chemistry?

( ) Yes

( ) Sometimes, depending on the content

( ) No

Justify your answer:

____

____

2. Do you find chemistry difficult to understand? Please explain.

____

____

____

3. Can you spot chemical reactions happening around you on a daily basis?

( )No ( ) Yes.

If you answered yes, name two:

____

____

4. Identify _ four products used in your home that contain chemical components.

____

____

5. Which of these products are considered acidic or basic?

____

____

6. When we have stomach pain, we take an antacid to make us feel better. Why?

7. According to your knowledge, explain why our body is a structure rich in chemistry.

8. What is your knowledge of organic and inorganic compounds?

9. Mark with an X those products that you consider to be organic.

( ) Medicines	( ) gasoline
( ) Oil	( ) Fat
( ) Carbon dioxide	( ) Detergent
( ) Alcohol	( ) Table salt
( ) Cooking oil	( ) Vinegar
( ) Water	( ) Butter
( ) Margarine	( ) Sodium bicarbonate
( ) Cooking gas	( ) Acetone

10. Mark with an X those compounds that you consider organic.

( ) CO_2	( ) CH_2OH
( ) NH_4OCN	( ) $MgCl$
( ) $CO(NH)_2$	( ) CH_3COOH
( ) $NaCl$	( ) $NaHCO_3$
( ) C_4H_{10}	( ) CH_3OCH_3
( ) $C_6H_{12}O_6$	( ) CH_4
( ) H_2O	( ) HCN
( ) CH_3CH_2OH	( ) HCl

11. In question 9, there are some products whose molecular formula can be found in the question 10. Identify these products and their molecular formulas.

APPENDIX III: Experimental lesson 1
PRACTICE 1-IDENTIFIED PHENOL IN PARACETAMOL, A MEDICINE USED TO TREAT DENGUE
INTRODUCTION

Dengue is considered to be one of the main public health problems in the world. It is caused by a virus and its vector is the Aedes aegypti mosquito. To reduce its transmission, it is necessary to combat the mosquito. In the event of contamination, appropriate medication should be used so that symptoms do not worsen.

One of the medications used to treat dengue is paracetamol, a drug with an active ingredient that makes it effective in combating pain and controlling temperature, as it belongs to the analgesic and antipyretic class. However, its inappropriate use can be harmful to the liver due to its toxicity.

Infection with different viruses results in symptoms ranging from high fever and muscle pain, which can worsen to bleeding caused by changes in blood clotting.

A person with dengue fever is already prone to bleeding, because the virus that transmits this disease attacks the platelets responsible for blood clotting. This bleeding symptom can be aggravated by the inappropriate use of drugs such as aspirin with the active ingredient acetylsalicylic acid, which has an anticoagulant effect, resulting in a decrease in blood clotting.

CHALLENGE

Imagine that a student has been infected with the dengue virus and in the school office there are two medicines available, manually labeled as aspirin and paracetamol, but it is suspected that the labels have been switched.

By studying organic fungi in chemistry, you will help the student to identify the correct medicine to use for dengue symptoms.

Firstly, the organic fungi present in these medicines were observed.

PROCEDURE

1 - Crush the tablets and place them in test tubes labeled A and B.
2 - Dilute the two samples in tube A and B with approximately 5 mL of methanol.
3 - Add 3 drops of 1% ferric chloride solution using a Pasteur pipette.
4 - Observe the reaction and color change in both test tubes. The drug that reacted is the one indicated for the treatment of dengue fever.

Below is the reaction for this experiment:

QUESTIONS

1 - Which organic fungi are present in paracetamol?

____

2 - What about aspirin? ____
3 - According to the results, which test tube contains the paracetamol sample, A or B?
4 - Based on the practice carried out and the observation of the reaction, ferric chloride is used to identify which organic function?
5 - Represent the physical states of each molecule (reactants and products) in the reaction.
6 - Show the structure of the solvent used.

REFERENCES

BRAZIL. Ministry of Health. National Health Surveillance Agency (ANVISA). 32 paracetamol.

COELHO, Giovanini Evelim. Dengue: current challenges. **Epidemiología e Servigos de Saúde**, v. 17, n. 3, p. 231-233, 2008.

BRAGA, Ima Aparecida; VALLE, Denise. Aedes aegypti: history of control in Brazil. **Epidemiología e servigos de saúde**, v. 16, n. 2, p. 113-118, 2007.

PAZINATO, M. S.; BRAIBANTE, H. T. S.; BRAIBANTE, M. E. F.; TREVISAN, M. C.; SILVA, G. S. A differentiated approach to teaching organic fungi through the theme of medicines. **Revista Química Nova na Escola**, v. 34, n° 1, p. 21-25, feb. 2012.

7 .4 APPENDIX IV: Experimental lesson 2

PRACTICE 2 - IDENTIFICATION OF ANTACIDS: SODIUM BICARBONATE AND ALUMINUM HYDROXIDE (PEPSAMAR) BY REACTION WITH HYDROCHLORIC ACID

INTRODUCTION

Antacids are basic medicines that react with the hydrochloric acid present in the gastric juice secreted by the cells of the stomach. As this is an over-the-counter drug, people abuse its use or self-medicate inappropriately when they believe that all drugs belonging to this pharmacological class have the same effects.

Aluminum hydroxide is a weak base that reacts with hydrochloric acid to form aluminum chloride and water, as shown in the reaction below:

$$Al(OH)_3 + 3HCl \rightarrow AlCl_3 + 3H_2O$$

Sodium bicarbonate is a basic salt and when it reacts with hydrochloric acid, sodium chloride (NaCl), known as table salt, water (H_2O) and carbon dioxide (CO_2) are formed, as shown in the reaction below:

$$NaHCO_3 + HCl \rightarrow NaCl + H_2O + CO_2$$

People with hypertension should avoid ingesting sodium chloride (NaCl) or substances that react with it, which is what happens with sodium bicarbonate.

CHALLENGE

In a practical lesson, the students were challenged to identify which chemical substance represents the salt sodium bicarbonate and which represents the base aluminum hydroxide. Both drugs have the same color characteristic, being white.

Your knowledge of chemistry will help you to identify which medicine a person with hypertension can use, in this case aluminum hydroxide, trade name Pepsamar, since sodium bicarbonate can intensify the problems related to your blood pressure.

From the reactions above, you can see that sodium bicarbonate reacts with acids and releases carbon dioxide (CO_2), which does not occur with the reaction of hydrochloric acid and aluminum hydroxide.

PROCEDURE

1 - Macerate the tablet so that the two drugs have the same
physical characteristics, as well as speeding up the reaction.

2 - Place both drugs in a test tube labeled C and D.

3 - Add approximately 5 mL of hydrochloric acid to both samples in tubes C and D.

4 - Observe the reaction through the release of gas.

QUESTIONS

1 - Based on the results, which drug is C?

---

2   - What about drug D?

3 - What medication can a person with hypertension use?

4 - Based on the experiment and knowledge of organic chemistry, can sodium bicarbonate be characterized to identify which organic function?

5 - Represent the reactions below by designating the physical states of each molecule and identifying the reactants and products.

REFERENCES

DE BARROS, H. C. M. Potassium and bicarbonate. **J Bras Nefrol**, v. 26, n. 3 - Supl 1, p. 23, 2004.

DE FREITAS, É. L. et al. Profile of antacid use by users of the UFMG university pharmacy, Belo Horizonte (MG). **Infarma-Ciências Farmacéuticas**, v. 18, n. 9/10, p. 36-40, 2013.

PAZINATO, M. S.; BRAIBANTE, H. T. S.; BRAIBANTE, M. E. F.; TREVISAN,M. C.; SILVA, G. S. A differentiated approach to teaching organic functions through the theme of medicines. **Revista Química Nova na Escola**, v. 34, n° 1, p. 21-25, feb. 2012.

APPENDIX V: Experimental lesson 3

PRACTICE 3: SIMILARITIES AND DIFFERENCES BETWEEN ALDEHYDES AND KETONES IN A PROPOSAL TO IDENTIFY FORMALDEHYDE

INTRODUCTION

Aldehydes and ketones have the carbonyl functional group:

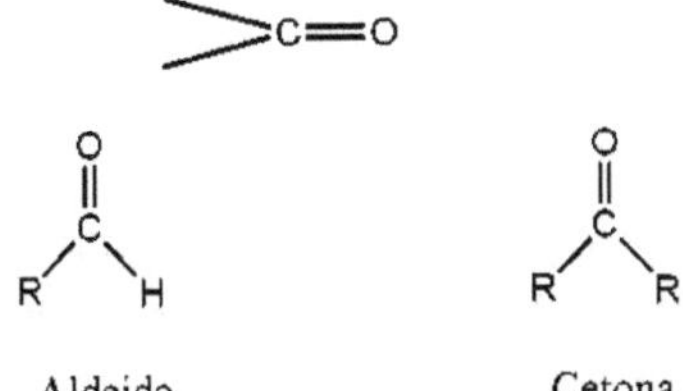

Aldeido   Cetona

   In aldehydes, the carbon atom of the carbonyl is linked to a hydrogen and an alkyl group represented by the letter R, which can be a carbon or a carbon chain; in ketones, the carbonyl is linked to two alkyl groups which can be the same or different.

   The characteristics of aldehydes and ketones are very similar, as their reactions occur at the carbonyl group. However, aldehydes are easy to oxidize by losing the hydrogen attached to the carbon atom of the carbonyl. Ketones, on the other hand, do not have this hydrogen, so they are more resistant to oxidation.

CHALLENGE

   Imagine that there are two unlabeled reagents in the school laboratory: formaldehyde and ketone. Using their knowledge of organic chemistry, the students will help the teacher to identify these reagents.

   Aldehydes and ketones can be differentiated using Tollens' reagent, a mixture of aqueous silver nitrate with sodium hydroxide and aqueous ammonia, as shown in the reactions below:

$$2\ AgNO_{3(aq)} + 2\ NaOH_{(aq)} \rightarrow Ag_2O_{(s)} + H_2O_{(l)} + 2\ NaNO_{3(aq)}$$

$$AgO_2 + 4\ NH_{3(aq)} + H_2O_{(l)} \rightarrow 2\ Ag(NH_3)_2OH_{(aq)}\ (\text{Reagente de Tollens})$$

In the reaction of aldehyde with Tollens' reagent, as the aldehyde oxidizes, there is a reduction of silver observed by the precipitation or formation of a silver mirror, according to the equation below:

$$\text{Aldeido} \quad + \quad 2\ Ag(NH_3)_2OH \quad \longrightarrow \quad 2\ Ag \quad + \quad \text{(R—COO}^-\,NH_4^+) \quad + \quad NH_3 \quad + \quad H_2O$$

precipitado de prata

Aldeido

The presence of a ketone instead of an aldehyde leads to a negative result in the presence of Tollens' reagent, as no precipitate forms because the ketone cannot be oxidized due to the fact that it has no hydrogen attached to the carbonyl group:

$$\text{Cetona} \quad + \quad 2\ Ag(NH_3)_2OH \quad \longrightarrow \quad \text{nenhuma reação}$$

Cetona

PROCEDURE

The two unlabeled reagents are labeled E and F for analysis.

1 - Separate one of the test tubes and add: 20 drops of silver nitrate solution, AgNO3 10% (m/v) and 20 drops of caustic soda solution, NaOH 10% (m/v). To the brown precipitate that forms, add ammonia solution, NH3 6.0 mol/L, drop by drop, until the precipitate dissolves.

2 - In another test tube, add two drops of the reagent E to be analyzed and dilute with 95% ethanol. Then add this ethanolic solution to the tube containing the Tollens reagent prepared in the previous section and observe.

3 - Repeat this procedure using reagent F. The formation of a silver precipitate or mirror in one of the two reagents will indicate the presence of a specific functional group.

QUESTIONS:

1 - Based on the results, which regent is the ketone and which is the formaldehyde?

2 - Why are aldehydes easy to oxidize and ketones not?

3 - Represent the aldehyde identification reaction by designating the physical states of each molecule and identifying the reactants and products.

REFERENCE

BARREIROS, André Luís Bacelar Silva; BARREIROS, Marizeth Libório. EXPERIMENTAL CARBOHYDRATES.

FUKUSHIMA, Ms André Rinaldi; DE ANDRADE, Ms Lucia Machado; LAMOLHA, Ms Marco Aurélio. Organic Chemistry Laboratory Workbook.

SOLOMONS, T. W. Graham; FRYHLE, Craig B. **Organic Chemistry**, vol. 1 and 2. 9 ed. LTC, 2009.

APPENDIX VI: Students' final questionnaire

Final questionnaire

Student's name: ____

1. Select the correct alternative. What is an organic compound?
a. Natural compounds, which come from nature.
b. Compounds that cannot be synthesized in laboratories.
c. They have chain-linked carbon atoms and/or carbon atoms linked directly to hydrogen.
d. They only have carbon and hydrogen atoms.

2. Below are substances from our daily lives. Which of the alternatives are considered to be organic compounds?
a. Medicines, cooking salt, alcohol.   c. Alcohol, cooking gas, water.

b. Medicines, cooking gas, alcohol.   d. Vinegar, butter, caustic soda
3. Which of the alternatives shows molecular formulas where all the compounds are organic?

a. CH_2OH - $NaCl$ - C_3H_6O c. HCl - CH_3CH_2OH - $C_6H_{12}O_6$
b. $NaHCO_3$ - CH_4 - $C_6H_{12}O_6$ d. C_3H_6O - CH_4 - CH_3COOH

4. The organic compounds: ethanol (ethyl alcohol), acetic acid (vinegar), butane (cooking gas) and propanone (acetone) belong to the organic fungi respectively:
a. Hydrocarbon, carboxylic acid, alcohol, ketone
b. Alcohol, hydrocarbon, carboxylic acid, ketone
c. Alcohol, carboxylic acid, hydrocarbon, ketone
d. Alcohol, carboxylic acid, ketone, hydrocarbon
5. Paracetamol is a drug with an active ingredient that makes it effective in combating pain and controlling temperature, as it belongs to the analgesic and antipyretic class. Which organic fungi are present in paracetamol?
a. Amine and phenol
b. Amide and phenol
c. Amine and alcohol
d. Amine and enol

6. Aspirin is a medicine with the active ingredient acetylsalicylic acid and is also used as an analgesic and antipyretic. Which organic compounds are present in aspirin?
a. Carboxylic acid and ester
b. Carboxylic acid and ketone
c. Alcohol and carboxylic acid
d. Carboxylic acid and ether

7. Codeine has active ingredients that are indicated for the treatment of severe pain, coughs and diarrhea. What functions are present in codeine?
a. Ether, ether, enol and amine
b. Ester, ether, alcohol and amine
c. Ether, ether, alcohol and amide
d. Ether, ether, alcohol and amine
8. Vitamin C is found in citrus fruits, vegetables and legumes. It benefits the appearance of the skin and the growth and repair of tissues. What functions does vitamin C perform?

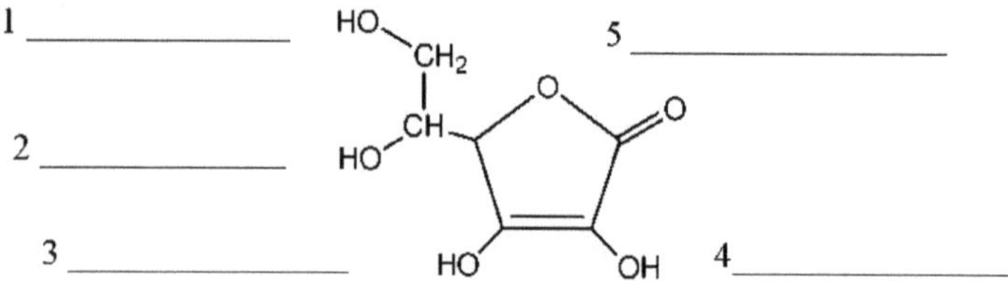

a. Alcohol, alcohol, enol, enol and ketone
b. Alcohol, alcohol, enol, enol and ether
c. Alcohol, alcohol, alcohol, alcohol and ether
d. Alcohol, alcohol, enol, enol and ester

9. When we use antacids, the hydrochloric acid (HCl) produced in excess by the stomach is neutralized. Below are two reactions using different antacids, aluminum hydroxide and sodium bicarbonate. Based on the products generated, explain which medicine would be more suitable for a hypertensive person, ALUMINIUM HYDROXIDE OR SODIUM BICARBONATE?

$$Al(OH)_3 + 3HCl \rightarrow AlCl_3 + 3H_2O$$

$$NaHCO_3 + HCl \rightarrow NaCl + H_2O + CO_2$$

10. Below are the structural formulas of two organic compounds: formaldehyde, popularly known as formaldehyde, which has preservative properties because it prevents the growth of microorganisms, and acetone, which is widely used as a nail polish remover. Check the INCORRECT alternative:

a. The first structure is formaldehyde
b. The organic function of the second structure is an aldehyde
c. Acetone can also be called propanone
d. Both structures have the carbonyl functional group